Living in a Community

Every living creature is part of a community of life. The smallest insect is a member of a plant and animal association whose members are constantly affecting one another's lives. Scientists call these communities **ecosystems**. An ecosystem is a naturally occurring collection of plants and animals that live in balance with each other and the soils and climate around them.

Take a look at your community. Are there plants and animals living around you that depend on one another? What plants and animals do *you* depend on? You can learn more about the interaction of plants and animals in your ecosystem by observing a specific area carefully.

A Community of Life In and Around a Tree

Find a tree that is a part of your living space. Mark a 5-foot (1.5-meter) length of adding-machine tape in 1-foot (30 cm) intervals. Stretch the tape from the ground up the trunk of the tree. Attach the top of the tape to the trunk with a tack or pushpin. This will not harm the tree. Now, using the record form on page 2, observe everything that is on the tree trunk—living and nonliving: Take notes to describe what happens as you watch your special area. List the things you see on the ground below. List the things you see in the tree above your tape.

Spend thirty minutes watching. Leave the tape in place and check the spot over a series of days to see if changes occur. Compare the observations that you have made. Are there plants and animals in your watch area that live in balance with each other? Write down any relationships that you observe.

Scientists call this kind of a standard plot a *transect*. They use observations of transects to help them learn about the relationships between plants and animals.

A Community of Life In and Around a Tree

Date _____   Time _____

List everything you see at each level. Draw some of the things if you wish.

Above 5 feet	
4 feet–5 feet (120–150 cm)	
3 feet–4 feet (90–120 cm)	
2 feet–3 feet (60–90 cm)	
1 foot –2 feet (30–60 cm)	
Ground–1 foot (0–30 cm)	
On the ground below the tree	

EMC 4125

R A I N F O R E S T

I magine that you are in a **rainforest**. The trees are so tall that they seem to touch the sky and block out most of the sunlight. Huge butterflies float through the air, and beetles as big as baseballs crawl on the ground. The air is damp. Rain drips off large leaves and tiny frogs cling to their tips. Their croaks join the cries of howler monkeys and scarlet macaws.

Rainforests are jungles that are found in the **tropical** regions of our world. Temperatures in the rainforest remain nearly the same all the time—night, day, summer, or winter. The average temperature in the rainforest is 75 degrees Fahrenheit (24 degrees Celsius). There is plenty of rain, so the humidity is high. The air feels moist and steamy. Tropical rainforests average over 79 inches of rain per year. A rainforest is known for its abundant growth and its wide variety of plant and animal life.

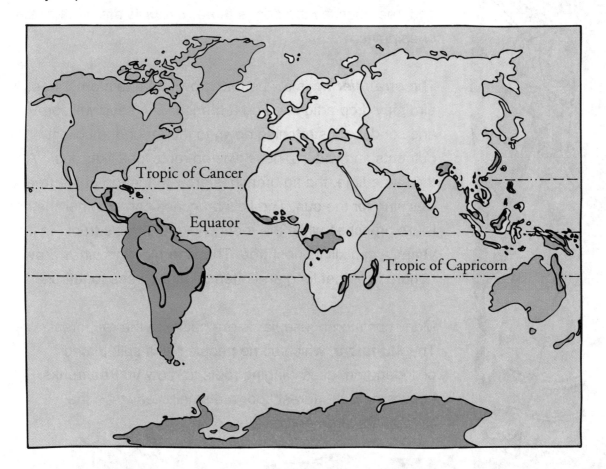

Look at the rainforests marked in dark green on the map. Notice that they are located near the **equator**, between the **Tropic of Cancer** and the **Tropic of Capricorn**. The climate for this region is called *tropical.*

PLANTS
of the rainforest

Giant trees of the rainforest stretch 150 feet in the air to capture the sun's energy, while underneath their spreading branches many other plants make use of the light that filters through.

Bromeliads are a kind of **Epiphyte** (e-pih-fite) — plants with no roots. They grow high up on other plants so they can get some sunlight. Their cupped leaves trap rainwater and store it in a tiny pool at the center of the plant. Unlike most plants that get nutrients from the soil, bromeliads have to get theirs from the rainwater. Mosquitoes breed in these leaf-pools, and tree frogs lay their eggs there. **Orchids** are another kind of epiphyte.

The **strangler fig** is a vine that grows on a rainforest tree. A bird may drop a fig seed on a branch. The seed will sprout and send roots stretching down to the ground to collect nutrients and water. These twisted roots look like ropes. At the same time, the fig branches grow to the top of the tree, reaching for the sun. The fig's branches can smother the host tree's branches while the roots choke the tree's trunk. In fifty years, it can kill its host tree. The strangler fig's vines grow hundreds of feet long and often tie many trees together.

Many common houseplants are native to the rainforest. The **Monstera**, which some people call a split-leafed philodendron, uses clinging roots to grow up tree trunks. Leaves near the forest floor are small. Leaves in the canopy are large and divided.

EMC 4125

The Kapok Tree

The *kapok* tree can be 150–200 feet tall. To support this great height, it develops giant roots, called *buttress roots*, that spread out beside the lower part of its trunk. It has large palm-like leaves. Its flowers, which bloom only during the flood season, produce hundreds of football-shaped pods. These pods are up to six inches long and filled with fibrous seeds. The pods are gathered after they fall on the ground or are cut from the tree with knives attached to long poles. The fruits are dried and the fiber inside is taken out. The fibers are bouyant (they float) and hold their shape. They are used to fill life preservers and as stuffing in some sleeping bags.

The kapok tree is a community in itself. Many plants and animals live in its sturdy branches. All of the animals listed in the word bank below can live in a Kapok tree. See if you can find them in the word search.

Word Bank

- bee
- boa constrictor
- coati
- iguana
- jaguar
- katydid
- kinkajou
- macaw
- ocelot
- parakeet
- postman butterfly
- silky anteater
- tamarin
- three-toed sloth
- toucan
- tree frog
- tree porcupine
- woolly monkey

```
a t r e e p o r c u p i n e b d e t
c f i j k p a r a k e e t l m h g h
n b o a c o n s t r i c t o r o p r
s q t x b s w z c s v a w y u r b e
i e t g f t i o c e l o t k h d e e
l k a j p m m n o r t q l i s o e t
k v m a c a w u a l x a d n w e y o
y c a z f n b g t j l i m k h l k e
a o r n t b v q i u z y x a p r s d
n y i i g u a n a a b d m j r w c s
t i n g j t n h k o m r p o l i q l
e u a w y t o u c a n v z u n b n o
a t f j n e d o h m g l p i q k c t
t r x s t r e e f r o g k u t v e h
e w b c y f d e z h n r g t f j o y
r a c k a l u i d j a g u a r y w s
l f m b q y e v p x z k a t y d i d
```

ANIMALS
of the rainforest

The rainforest contains hundreds of varieties of insects, birds, and animals. But if you walked into the forest you would have a hard time seeing many of them. The insects would be the easiest to spot. The rainforest is home to half of the world's *insects*! Many insects would be on the forest floor in the dark shade among the leaves, twigs, and rotting wood. Other rainforest creatures live in the high branches of the trees where fruit and new leaves are in abundance. There is also safety there, away from carnivores like the *ocelot* and the *jaguar.*

The *flying gecko* lives high in the rainforest tress. It doesn't really fly. It has flaps of skin on either side of its body that act as parachutes so that it can leap from tree to tree. Geckos are active mainly at night. Their excellent eyesight and hearing help them find insects. Ridges of scales under the gecko's feet enable it to cling to slippery surfaces and even run upside down along branches. Female geckos lay one or two eggs with soft sticky shells. They hide these eggs under a piece of tree bark.

EMC 4125

The bright orange and black *tiger centipede* scurries about the floor of the rainforest. This giant centipede is almost ten inches (25.5 cm) long. It feeds mainly on insects and spiders, but it can also catch small toads, snakes, and mammals. The centipedes prefer the cool damp nights and hide under leaves and logs during the day. Each centipede has a long, jointed body with as many as 23 pairs of legs. Two long antennae help it feel its way around. Two large legs at the end of the abdomen hold the centipede's prey still, while it injects poison with its large claws.

Flocks of colorful *toucans* hop through branches high in the rainforest and fly from tree to tree in search of their favorite food, passion fruit. They do not live in nests like many birds. Instead they live in holes in the trunks of the giant rainforest trees. The toucan's large bill looks heavy, but it is made of lightweight keratin— like your fingernails—and it is hollow. The serrated edges of the bill allow toucans to bite off chunks of fruit easily. Because its bill is so long, the toucan can reach berries and seeds on branches that are too thin for it to stand on. Once it has plucked a tasty morsel, the toucan tosses back its head and flicks the food down its throat.

LAYERS
in the rainforest

Plants in the rainforest grow in layers, depending on the amount of sunlight they need. Each layer gets less light than the layer above it. Rainforest animals live in the layer, or layers, where they can find food and shelter. Scientists divide the rainforest into four layers.

The top layer of the rainforest is called the **emergent layer**. Here, where giant trees tower about 150 feet above the ground, breezes blow and the sunlight is hot and glaring. It feels much less wet here, too. **Howler monkeys (1)** set up a ferocious roar, always wary of the huge **harpy eagle (2)**. Smaller animals visit the emergent layer as well: **lizards**, the **woolly mouse opossum (3)**, and the **spear-nosed bat.**

The **canopy**, the tops of medium-sized trees more than 60 feet above the ground, is the next layer of the rainforest. The dense green leaves of the canopy is where most of the photosynthesis (the food-making process of green plants) takes place. More animals live in the canopy than in any other part of the rainforest. **Three-toed sloths (4)** and **silky anteaters** crawl among the branches. **Macaws, parakeets, parrots,** and **toucans (6)** nest and perch. **Flying frogs** and beautiful **butterflies (5)** glide from limb to limb. This is a layer of warm, bright color.

The shady light of the **understory** is home to small trees and shrubs. Mosses, ferns, and orchids grow in the filtered light. This is the home of **mosquitoes** and **tree frogs (7)**. **Tree boas (8)** and other snakes slither among the leaves and vines. The largest rainforest predator, the **jaguar (9)**, hides on a branch in the dappled light.

On the dark **forest floor**, many large mammals, such as the **giant anteater (10)** and the **tapir (11)** search for food. **Ants, termites, beetles (12)**, and **scorpions** scurry through the rotting leaf litter. The **caiman**, an alligator-like reptile, waits along the river bank for unsuspecting prey.

EMC 4125

Make a Rainforest Terrarium

Materials:

- large bottle with a wide mouth or glass aquarium
- gravel
- sand
- peat moss
- small pan of water
- small plants: mosses, ferns, begonias, houseplants
- small tree frog, turtle, or salamander

Steps to Follow:

1. Mix equal amounts of gravel, sand, and peat moss for the soil of your terrarium. This will give you good drainage and allow air in the soil.

2. Put the soil in your terrarium. Make sure that the surface of your soil is at a slant (three or four inches thick at one end, slanting to the level of your pond [see step 3] at the other end).

3. Bury the pan of water at the low end. (This is the "pond" and will provide the base for your water cycle.)

4. Put in your small plants.

5. Add an animal if you wish.

6. Cover the top of your aquarium with a lid or piece of glass. If you have added an animal, you must allow some air into the terrarium. Punch holes in the lid or adjust the glass to leave a small opening. The water in the pond will provide enough moisture for the plants. Your rainforest will need sunlight.

Feed your animals sparingly and remove uneaten food or it will decay. Frogs and salamanders will eat bits of chopped meat, earthworms, and flies. Turtles eat bits of hard-boiled egg, lettuce, berries, and turtle food from the pet store.

The rainforest is enclosed much like your terrarium. In the rainforest, the canopy layer traps the heat and moisture, helping to make more "rain."

EMC 4125

D E S E R T

Imagine that you are in the **desert.** The sand beneath your feet is warm, even though it is early morning. There are few plants in sight. You can see a thorny saguaro cactus standing guard as a horned lizard rests on a flat rock nearby. It is quiet. A hawk glides overhead as you move on. Sand crickets and beetles dart away from a roadrunner looking for its breakfast.

At first glance, the desert looks like a place where there is very little life, but a closer look proves this wrong. There are over 5,000 species of animals that live in deserts.

Actually there are many different types of deserts. Any place that receives less than 10 inches of rain per year can be classified as a desert. So both **Death Valley** in California and **Siberia** in Russia are deserts. Deserts can be hot or cold, but most often we think of deserts as areas with very hot days.

Both the animals and the plants in the desert have adapted to the heat and lack of water. Seeds have tough seed coats and remain dormant until there is enough water to sprout. Trees have few leaves and large root systems. Animals are active mainly at dawn and dusk, or are nocturnal, hunting at night and sleeping during the day.

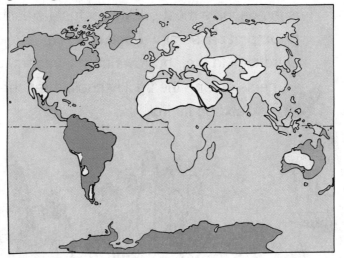

Look at the deserts marked in beige on the map. Which of the deserts do you think would be hot deserts? Which do you think would be cold deserts?

The life in one desert may be very different from that of another desert. The plants and animals described in this book are found in the **North American deserts**.

PLANTS
of the desert

The plants of the desert are good at surviving drought. They must be able to go without water for long periods. Many of the plants are **succulents**. Succulents store water in their leaves in order to make it through droughts. During rainstorms, moisture is carried to spongy storage tissues inside the leaves and stems. The surface of the plant is often covered with a layer of wax or a blanket of fine hairs which keep the water from evaporating. Other desert plants with green leaves have very small leaves that reduce water loss. Desert plants often have deep tap roots that draw water from far below the surface. Many have extensive roots close to the surface that absorb moisture on the ground when rain does fall.

The **boojum tree** looks like an upside-down carrot. It stores water in its cone-shaped trunk. When it rains, leaves sprout all over it. The leaves last for a few weeks, and then their stems turn into thorns.

A **cholla cactus** has spines as sharp as needles and no leaves. It is the desert's most prickly cactus. Its easily detached spines seem to jump at passersby. Pack rats pile up cholla stems to protect the entrances to their burrows from invaders. Cactus wrens build their nests among the prickly cholla branches and ground squirrels feed on its fruit.

EMC 4125

A **creosote bush** is a wispy shrub with scraggly branches two to five feet high. It helps to keep desert soil from eroding and provides shade and food for desert animals. The sand mounds that form around it are used as home-building sites by desert animals such as ground squirrels, kangaroo rats, lizards, snakes, and toads.

The long-branched **ocotillo plant** is covered with leaves following a period of rainfall. As soon as the wet period is ended, the leaves are shed. An armor of tough cells then protects the plant from water loss. When rainfall is scattered, several crops of leaves may appear in a year. The long thorns that remain on its branches are actually the stem and midrib of its leaves.

The **Joshua tree** has sword-like leaves that grow in bunches at the ends of its branches. One area of the California desert where the Joshua trees grow has been declared a national monument.

Ocotillo

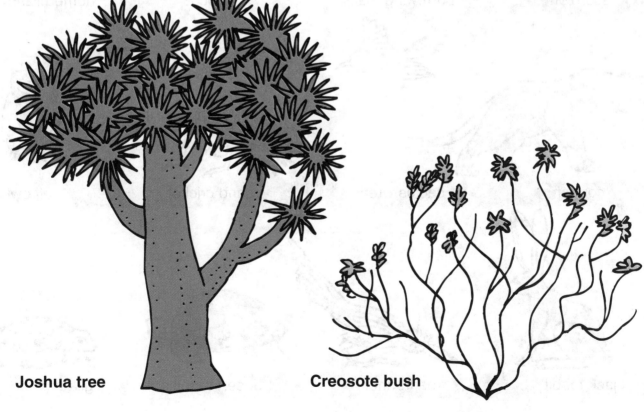

Joshua tree

Creosote bush

ANIMALS
of the desert

It is not easy to live in the desert. Yet many different kinds of animals are successful in this harsh environment. Here are some animals of the North American desert. Classify these animals by writing the names in the appropriate boxes on the next page.

kit fox sidewinder rattlesnake canyon tree frog tarantula

scorpion burrowing toad bobcat horned lizard

cougar cactus wren sand cricket elf owl

jack rabbit roadrunner desert tortoise gila monster

EMC 4125

ANIMALS
of the desert

REPTILES

MAMMALS

AMPHIBIANS

BIRDS

INVERTEBRATES

LAYERS
in the desert

Like the giant kapok tree of the rainforest, the **saguaro cactus** is a community within itself. The saguaro grows only in the **Sonoran Desert,** which is located in parts of Arizona, California, and Mexico. The saguaro can be fifty feet tall, weigh several tons, and live for two hundred years. Its sharp spines protect it and the accordion-like pleats in its skin expand in the rain, storing extra water for long, dry days. A huge network of shallow roots spreads in all directions to absorb as much rainwater as possible.

The **Gila woodpecker** pecks a hole high up in the trunk of the saguaro and builds a nest for its eggs. The cactus flesh forms a hard lining around the nests. When the cactus dies and decomposes, these hollow forms are left. Indians call them "saguaro boots" and use them to carry food.

The **elf owl** also lives in the saguaro. The thick lining of the cactus and the moisture stored inside make the owl's nest cool during the heat of the day. **Harris' hawks** build their nests in the notch between the trunk and the arm of the saguaro.

The saguaro blossoms in May. Large white flowers with yellow centers open on the top of the cactus. Each flower opens only once, at night, and closes the following afternoon. **Long-tongued bats** and **doves** drink the flower nectar. When fruit forms at the base of the blossoms in June, birds, animals, and insects enjoy its sweet, red pulp.

After many years, when the saguaro dies, the outer flesh falls to the ground. The woody saguaro ribs may be left standing, like a tall broom. Still, the cactus is home for a whole host of creatures. **Termites**, **centipedes**, **spiders**, **lizards**, **mice**, and **snakes** are among the animals who move into the dead saguaro.

The illustration on the next page shows the saguaro community.

EMC 4125

What Do a
Teddy Bear and a Panda
Have in Common?

They are both desert plants! A lot of desert plants have imaginative names. When you hear them, you might get a funny picture in your mind. Draw what the names suggest these plants might look like. (Go to the Answer Key to see what they really look like.)

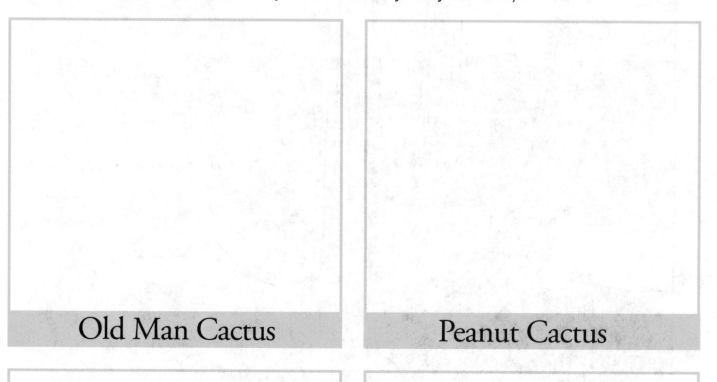

| Old Man Cactus | Peanut Cactus |
| Rainbow Cactus | Bunny Ears |

EMC 4125

Make a Desert Dish Garden

Materials:

- a shallow dish (size will depend on number of plants you have)
- several small cactus plants
- potting soil
- sand (not beach sand)
- gravel
- measuring cup
- spoon
- old towel or folded newspaper for handling the cactus plants

Steps to Follow:

1. Mix equal parts potting soil and sand.

2. Put a layer of gravel in the bottom of your dish for good drainage.

3. Spoon in the soil/sand mixture.

4. Make a hole for each cactus.

5. Put the cacti in the holes and gently press the soil to make it firm.

6. You can put pebbles, rocks, or other decorative objects on top of the soil.

Caring for Your Desert Dish Garden:

1. Water: The best way to water your dish garden is to set it in a sink or pan so that water just covers the dish. You will see bubbles rising. When the bubbles stop, take out the dish. Prop up one side of the dish garden so that extra water can drain away. Water when the soil is quite dry.

2. Light: Most cacti will do well near a south-facing or west-facing window.

3. Resting Period: Cacti like a resting period during the winter. Move them to a cool room and water only about every month.

4. Feeding: When your cacti are blooming, give them liquid tomato fertilizer once a month.

Imagine that you are standing on the shore watching the powerful waves pound the rocks. Here, anchored to the rocks and clinging to the seaweed, are a wide variety of fish and invertebrates. Tiny crabs scurry beneath rocks, and bright anemones wave their tentacles. This is the intertidal zone—the place where the tides flow in and out. As you move deeper into the water, huge blades of kelp float beside you. A stingray "flies" by as you descend into an underwater forest. You may pass a reef as you head out to the open sea. Watch as the marlin and tuna pursue schools of smaller fish like snapper and herring. You are still moving down. Here in the darkness of the deep sea, fish like the bristle mouth and gulper eel are scavengers.

The ocean is a collection of many different ecosystems. The **intertidal zone**, the **kelp forest**, the **coral reef**, the **open sea**, and the **ocean bottom** each form their own community of interdependent plants and animals.

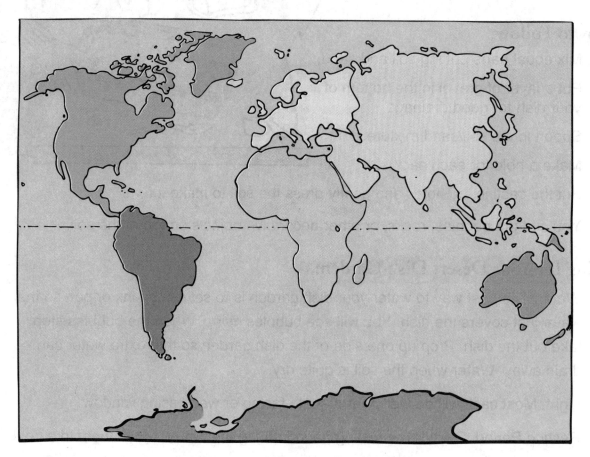

Notice how much more water there is than land. Two-thirds of the Earth is covered with water. Perhaps our planet should have been named "Oceana," rather than "Earth."

EMC 4125

The Kelp Forest

One of the specialized ecosystems of the ocean is the **kelp forest**. You will find kelp forests along coastlines where there are hard surfaces for the kelp to attach to, plenty of nutrients, moderate water motion, and clear, cool (70 degrees Fahrenheit or lower) ocean water.

Kelp is a name for a group of large seaweeds, or **giant brown algae,** that sometimes grow over 100 feet tall. There are twenty different species of kelp that grow along the coast of California alone.

There are three parts of a kelp plant: the holdfast, the stipe, and the blades.

The **holdfast** secures the kelp to the rock. It is root-like, but it is not actually a root: It does not gather water and nutrients and send them up to the kelp; it simply holds fast to a rock and secures the kelp so that it will not be swept away by the ocean tides. The holdfast becomes the home for many small sea creatures—crabs, snails, brittle stars, and worms.

The **stipe** is like a stem. It is tough but flexible. It is the anchor line—a vital connection between the holdfast and the blades. Food moves through the stipe to the bottom. The stipe has **floats**, hollow bumps filled with air that provide buoyancy. The floats pull the blades to the surface where they can get more sunlight. One type of kelp, the giant kelp, has one float for each blade. Another type, the bull kelp, has one big float for the whole plant.

The **blade** looks like a leaf. It absorbs water, carbon dioxide, and other chemicals. Then it uses energy from sunlight to convert those elements into oxygen and food for the plant. This process is called photo-synthesis and is an important function of all green plants. The blade also produces spores which, in turn, produce new kelp plants.

blade

float

stipe

holdfast

ANIMALS
of the kelp forest

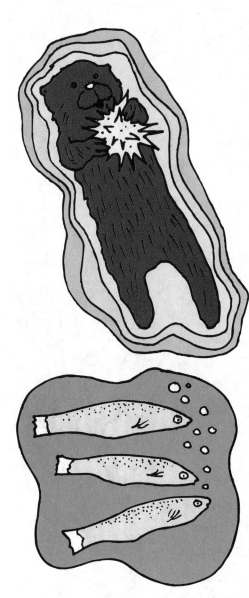

Sea Otter

The sea otter is found along the coast of the Pacific Ocean. It doesn't have a blubber layer like many other sea mammals do, so it must eat 25% of its body weight every day to get enough energy to maintain its body heat in the cold Pacific waters. After diving to the bottom to find an abalone or sea urchin, it anchors itself in the blades of kelp and uses a stone to break the hard shell of its meal. Sea otters weigh as much as 100 pounds (45 kg) and are good swimmers, resurfacing about every two minutes for air.

The sea otter has the thickest fur of any mammal. The desirability of this for clothing nearly caused the otter's extinction. Once the otter was considered an endangered species, but protection from hunters in recent years has helped the sea otter population to increase.

Señoritas

Señoritas are small, yellow, cigar-shaped fish. They swim among the stipes of the kelp forest and pick invertebrates off the kelp with their protruding buck teeth.They even pick tiny creatures from the backs of other fishes. If they are frightened, they dart to the ocean floor and hide by burrowing in the sand. At night, they rest in the sand with only their heads exposed.

Opossum Shrimp

Opossum shrimp live in the canopy of the kelp forest. They are tiny transparent creatures that are eaten by squid and fish. They carry their young in pouches. During the day, they hide among the blades, and at night they swim in search of the small crustaceans that they eat.

EMC 4125

Ribbon Worms

There are 600 kinds of worms that live in and around the holdfast of the kelp plant. The ribbon worm is about 8 inches (20 cm) long normally, but can extend itself to a yard (meter). It lives in a parchment-like tube anchored among the algae. It gathers food using a long tongue-like appendage, called a proboscis, that shoots out, secreting a sticky mucus to help catch its prey. Ribbon worms are night carnivores, eating other worms, small fish, mollusks, and small crabs.

Sea Stars

The underside of a sea star (sometimes called starfish) is covered with hundreds of tube feet, powered by a fluid-filled hydraulic system. The tube feet act like suction cups, pushing and pulling to move the star. Sea stars seem rigid, but the flexibility of their arms is shown when a wave flips one over. The arms curl under, while the tiny tube feet get a grip on a rock. Slowly the animal rights itself. A sea star eats mollusks, such as mussels and clams. Its mouth is located at the center of its body on the underside. Sea stars can grow a new arm if one is damaged or cut off.

Sea Star Cousins

Some sea star relatives look nothing like a sea star. See if you can match each description with the correct picture.

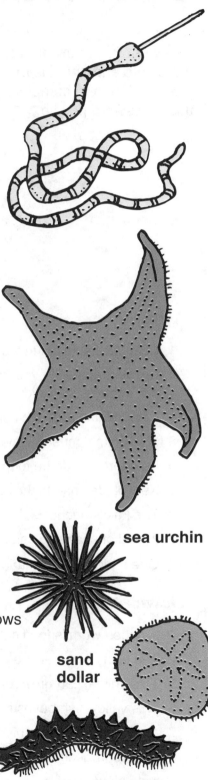

sea urchin

sand dollar

sea cucumber

1. This animal starts out as a star. Its rays never grow, but the central part of its body grows into a long tube. It still has five rows of little tube feet. All five rows end up on one side of the body.

2. This animal starts out as a star. Its rays bend up and grow together, but then flatten down. It grows short hair-like spines.

3. This animal starts out as a star, but its rays fold up over its top and grow together. Long spines grow and make it look like a pincushion.

Ocean Crossword

Use what you know about the plants and animals of the kelp forest to complete this crossword puzzle.

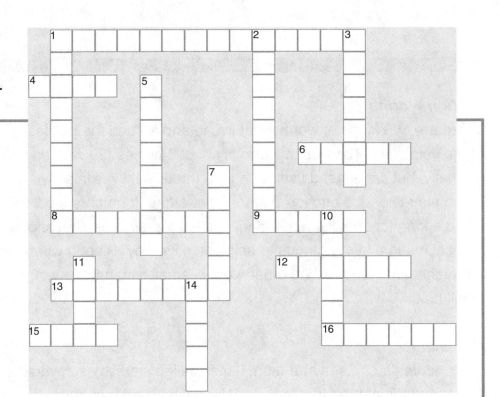

Across

1. Kelp uses sunlight in the process of _____.
4. Unlike the ones in your garden, this ribbon _____ can stretch to more than a meter long.
6. The leaf-like part of a kelp plant
8. This animal must eat over 25% of its body weight daily.
9. Kelp forests can be found along the _____.
12. Kelp blades rise to the surface because of their _____.
13. The kelp's anchor
15. A kind of seaweed belonging to the brown algae family
16. One variety of this marine animal carries its babies in a pouch.

Down

1. A ribbon worm's food catcher
2. The spiny sea star uses a _____ system to move.
3. This animal can grow a new arm it if looses one.
5. This yellow fish buries itself in the sand when disturbed.
7. A community of kelp plants is called a _____.
10. Kelp reproduce with _____.
11. Kelp plants prefer _____ water.
14. The stem-like part of a kelp plant

Word Box

blade

coast

cool

floats

forest

holdfast

hydraulic

kelp

photosynthesis

proboscis

sea otter

sea star

senorita

shrimp

spores

stipe

worm

EMC 4125

LAYERS
in the kelp forest

Even in the kelp forest, animals and plants seem to live in layers. Cut out the animals at the bottom of this page and put them in the right layer.

Near the surface, the kelp canopy floats in the water. The juvenile stages of many invertebrates like *hydroids* and *bryozoans* hide among the blades. **Sea otters** anchor themselves while they sleep by wrapping up in the kelp. They balance food, such as abalone, on their stomachs and use a rock to crack the shells open.

Along the stipes, the **turban snails** move up and down. Tiny drifting plants and animals in plankton float by. Here there are many slender fish like the **señoritas**.

Inside the holdfast there is another community of sea creatures. There are **brittle stars, worms**, and **hermit crabs**. These creatures nibble the stubby branches of the holdfast and remain in relative safety inside.

Create an Aquarium Ecosystem

A freshwater aquarium is the simplest. Saltwater aquariums are expensive and quite difficult to maintain.

Materials:

- a container—large-mouth gallon jar or a commercial aquarium

- fish—buy small fish at a pet shop or catch small bass, crappies, bluegills, suckers, and or catfish from a lake or pond. Be sure to get some of the water in which the fish are swimming when you catch them.

- gravel

- plants—purchased at a pet shop or from the same pond or river as the fish

- glass cover for the aquarium

- newspaper

- fish food

Steps to Follow:

1. Place two inches of gravel in your tank.

2. Lay a piece of folded newspaper on the gravel and carefully fill the aquarium with water. Then remove the newspaper. *Note: If you fill the aquarium with tap water, let it stand a few days before you put in any fish.*

3. Put in the plants.

4. Add the fish. (If you caught the fish, place the fish and its open water container in your aquarium for fifteen minutes before emptying the fish into the aquarium. If there is a difference in water temperature, the fish will be introduced to that change gradually.)

5. Cover the aquarium with a piece of glass. This will keep the fish from jumping out and keep the water from evaporating too rapidly.

6. Place the aquarium near a window so that it gets a sufficient amount of light.

EMC 4125

Comparing "TREES" of Different Ecosystems

The kapok tree, the saguaro cactus, and the giant kelp have many similarities, even though they are each found in very different ecosystems. Find the information needed to complete the chart below.

	Kapok Tree	Saguaro Cactus	Giant Kelp
Where found			
"Root" system			
Trunk			
"Leaves"			
Life they support			

- Brazilian rainforest
- blades
- cool oceans
- large buttress roots
- elf owl, woodpecker, scorpion

- sharp spines
- Sonoran Desert
- palm-like leaves
- toucan, anteater, tree frog
- 150-200 feet tall

- huge network of shallow roots
- stem-like stipe
- sea star, señorita, turban snail
- tough, accordion-pleated
- holdfast; not really roots

Human Impact on Nature's Ecosystems

Rainforests Are Disappearing

The rainforests are the source of much of the Earth's oxygen. New medicines and other valuable products are produced from plants and animals discovered in rainforests.

Between fifty and one hundred acres of rainforest are destroyed every minute as people clear land throughout the world. At the current rate of deforestation, there will be no tropical forest left within one hundred years.

- Nearly forty percent of the Amazon rainforest destroyed in Brazil was cleared for cattle ranching.

- Production of oil threatens large areas of tropical rainforest. Major oil production companies are all pushing their way deep into tropical forests in search of "black gold."

- Tropical hardwoods like mahogany and teak are valued for both aesthetic and practical reasons. Americans spend $2 billion a year on tropical wood products.

Deserts Are Changing

The desert offers many treasures to people. Native plants provide livestock forage and raw materials for drugs, fibers, dyes, and edible fruits. The desert offers year-round dry, clear air and plenty of space. Recently people have taken advantage of this and built suburbs and brand-new cities. Thousands of water wells have been drilled to bring water to the developments and to people, who are not drought resistant. The most desired riches of the desert are its deposits of oil and valuable minerals.

- Many of the deep wells being drilled today in deserts around the world are tapping water that can never be replaced.

- Failure to understand the terrible effects of overgrazing on desert vegetation has led to the collapse of civilizations more than all other factors together, including war, conquest, and pestilence.

- Over nine million barrels of oil flow daily from wells in the world's deserts.

EMC 4125

Oceans Are Being Polluted

Oceangoing vessels discard 14 billion pounds of cargo and crew wastes every year. While some litter sinks or rapidly decays, plastics have a life span of hundreds of years. Simple fishing line that is lost or thrown overboard is lethal for sea turtles and birds. Plastic trash, particularly six-pack rings, bags, and nets can trap fish and birds and strangle them as they try to break free. Even great whales are victims. Several have been found dead with plastic bags and sheeting in their stomachs.

- Large amount of fertilizers, which contain nitrogen, are washed into the sea by rivers carrying soil eroded from the land.

- Sewage pollution, warm water temperatures, and removal of lobsters and abalones by humans have destroyed many California kelp forests.

- Oil spills have killed thousands of marine mammals and sea birds.

Take a moment to think...

How would you be affected if rainforests did not exist, if desert ecosystems were destroyed, and the world's oceans were severely polluted?

There are organizations that are concerned about the future of our planet's ecosystems. You can write to them for information and to learn how you can help. Here are the names and addresses of a few of these groups:

Friends of the Earth: 1025 Vermont Ave., Suite 300, Washington, D.C. 20005

World Wide Web address: http://www.essential.org/foe.html

E-Mail: foedc@igc.apc.org

Friends of the Earth has an action packet for kids called *The Home Planet.* It is available at a cost of $5.00.

Nature Conservancy: to request information about these programs—*Adopt-an-Acre* (of rainforest) or *Rescue-a-Reef,* call 1-800-84-adopt; *Adopt-a-Bison*, call 1-800-628-6860.

Sierra Club: for a kid's newletter containing general environmental information, write Information Center, Sierra Club, 730 Polk Street, San Francisco, CA 94109.

Hiding in the Rainforest

How many rainforest animals can you find in this picture?

Playing It Cool in the Desert

How many lizards can you find in this picture?

Counting by 2's in the Ocean

Can you find 2 sea otters, 4 señoritas, 6 turban snails, and 8 sea stars?

EMC 4125